WHAT DID YOU FIND OUT?
Reporting Conclusions

Barbara A. Somervill

The Rosen Publishing Group's
PowerKids Press™
New York

For Maggie, Kate, and Joan

Published in 2007 by The Rosen Publishing Group, Inc.
29 East 21st Street, New York, NY 10010

First Edition

Editor: Joanne Randolph
Book Design: Elana Davidian
Layout Design: Julio Gil

Photo Credits: Cover © Larry Williams/Corbis; p. 4 © Steve Chenn/Corbis; p. 7 © www.istockphoto.com/Paul Cowan; p. 8 © JLP/Jose Luis Pelaez/zefa/Corbis; p. 11 © Meiko Arquillos/zefa/Corbis; p. 12 © www.istockphoto.com/Christopher Messer; p. 16 © LWA-Dann Tardif/Corbis; p. 19 © Layne Kennedy/Corbis; p. 20 © Tom Stewart/Corbis.

Library of Congress Cataloging-in-Publication Data

Somervill, Barbara A.
   What did you find out? : reporting conclusions / Barbara A. Somervill. — 1st ed.
      p. cm. — (Think like a scientist)
   Includes bibliographical references and index.
   ISBN 1-4042-3486-1 (lib. bdg.) — ISBN 1-4042-2195-6 (pbk.)
   1. Science—Methodology—Juvenile literature. 2. Science—Experiments—Juvenile literature. 3. Observation (Scientific method)—Juvenile literature. 4. Communication in science—Juvenile literature. I. Title. II. Series.
   Q175.2.S665 2007
   507.2—dc22
                                    2005035728

Manufactured in the United States of America

Contents

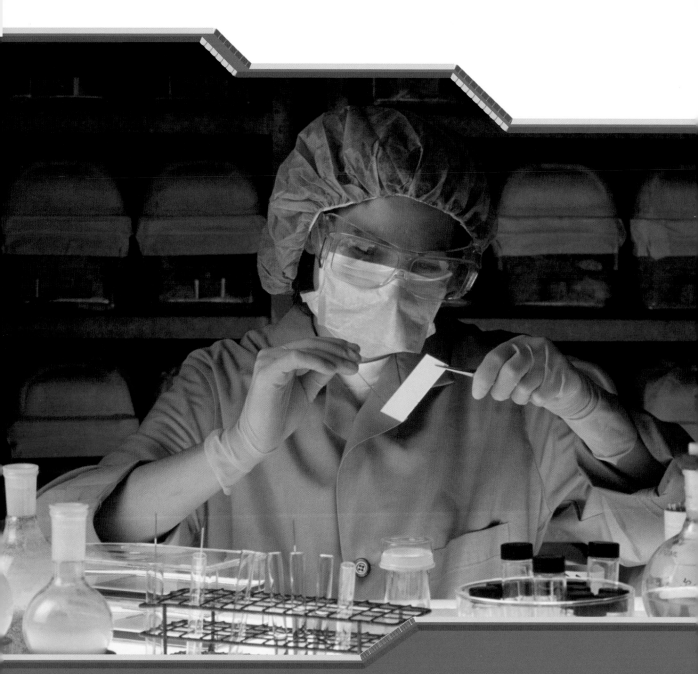

This scientist is using litmus paper in her work. As she tests the different liquids, she will collect information and record the results. She will want to report her final conclusions when she is done.

The Scientific Method

Every science experiment starts out with an idea. In science class, you might have learned to test for **acids** using **litmus** paper. One of the acids you probably tested was lemon juice. Let's say you then decided you want to test other foods at home to decide if they are acidic, **alkaline**, or **neutral**. That idea is the basis for your science experiment.

As with all science experiments, this experiment must follow the scientific method. You must start by making a guess or **hypothesis**. Your hypothesis is that most foods are acidic. Then you need to run an experiment and collect **information**. Next you need to decide what your results mean. The decision you make is a **valid** conclusion, or a statement based on facts.

What Is a Valid Conclusion?

Reaching a valid conclusion is the object of your experiment. You want to look at all the information you collect and decide what it means. You must ask yourself questions. What did you hope to prove? What does the **data** tell you? Did you prove your hypothesis?

Write your answers in a journal. When based on facts, these answers are valid conclusions.

Here is an example of a conclusion drawn from facts. You test lemon juice, orange juice, grapefruit juice, and lime juice and find they are all acidic. You know that the fruits you tested are citrus fruits. You conclude that all citrus fruits, including tangerines, have acid.

All the citrus fruits you tested were acidic. To use that information to say all fruits are acidic would not be a valid conclusion, though. To draw a conclusion about all fruits, you would have to test many more fruits.

It is important to take good notes about what happens during your experiment. Include important facts in your report. Be sure to use simple language so that people can understand your conclusions.

Writing Up Your Report

After doing an experiment, you must **communicate** your conclusions. A report is one way to do this. A report explains the experiment. It should begin with your original idea and what you hoped to prove. Be sure to include a list of supplies, your activities, and any data collected during the experiment. Your report should end with a **summary** of your experiment. This summary should include the conclusions you formed from doing the experiment.

Be sure to keep your report simple and use clear language. You want to make sure other people can understand what your experiment was about and what you found out by doing it. If you leave out important information, the reader might not understand how you reached your conclusions.

Answering the Five Ws

Your report should be much like a newspaper article. A good newspaper article tries to answer the five Ws, which are *who*, *what*, *where*, *when*, and *why*. If you do this in a science experiment, the answers should help you form correct conclusions.

The *who* in your experiment is you, the scientist. The *what* includes what you wanted to prove, what actions you took, and what you learned.

The place where your ran the experiment and how long each part took are the *where* and *when*. Finally your report must address the *why*. Why did you do this experiment? Why should others want to know what you found out? Why did you come to the conclusions you did?

This girl is testing foods in her kitchen. When she reports her conclusions, she will need to be sure to list exactly which foods she tested and what the results were.

Cranberries are an acidic food. Acidic foods often have a sour taste, as do these cranberries. Testing cranberries provides a positive result for your experiment.

Positive and Negative Results

In a science experiment, you will collect both positive and negative information. Positive data agrees with your basic idea. Negative data is information that disagrees with your ideas. By describing both types of results, you will present a balanced report.

In your experiment your hypothesis states that most foods are acidic. You test foods with litmus paper and find that lemons, corn, and cranberries are acidic foods. Baking soda, egg whites, and cooked spinach are alkaline. You find that water and soy milk are neutral. Finding foods that are not acidic is a negative result.

Consider the results you found and then come to a conclusion. You decide that not every food is acidic. That is a valid conclusion.

Charts, Graphs, and Diagrams

Words are not the only way to communicate the conclusions you reach by doing your experiment. Charts, **graphs**, and **diagrams** present your thoughts in picture form. These **visual** aids help you **organize** data and compare results.

It will help other people understand your graphic aids if you give each one a title. Label key parts of the picture or diagram. Add a caption, or explanation, under the picture to tell readers what they are looking at.

In your food experiment, you did tests and recorded the results. You can use graphs, charts, and diagrams to show people the results you used to form your conclusions.

Acidic, Alkaline, and Neutral Foods

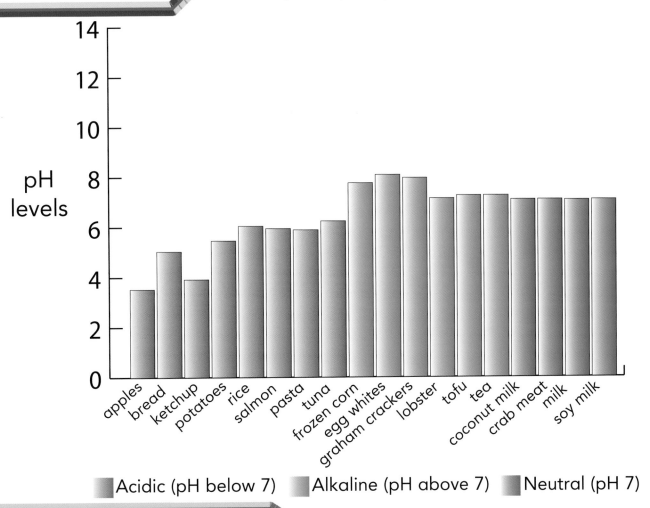

pH levels

Acidic (pH below 7)   Alkaline (pH above 7)   Neutral (pH 7)

This bar graph shows which foods are acidic, alkaline, or neutral. The acidic foods are in red, the alkaline ones are in blue. Neutral foods are in green. A pH level is a scale that tells you how acidic something is.

Creating a model is one way to report your conclusions. A model is a way to show people what you have learned. Here these women are creating a working model of a volcano.

Photos and Models

Photos and models are excellent aids for many science projects. Photos and models help you report your conclusions about your experiment to others. In the food experiment, color photos might be used to show the results you got by testing several foods, such as milk, lemon juice, baking soda, and egg whites. You could photograph the food with the litmus paper you used in front of the item. The photos help you show how you came to the conclusions you reached.

Some experiments can often be explained by using models. For example, a working model of a **volcano** can give information or explain how volcanoes work. It also draws interest and brings people over to learn more about your project.

Making Your Display

A display, poster, or presentation board shows people what you found out. Do not make a huge display that you cannot carry easily. A good size is about 3 feet (1 m) wide by 3 feet (1 m) tall. Make sure that any printed information can be read from 3 to 5 feet (1–1.5 m) away.

If you join two or three pieces of thick poster board, your display can stand on its own. Explain your experiment on the left. Describe the basic idea, the hypothesis, and the steps you followed. In the middle display pictures, graphs, and diagrams. On the right give your results and a conclusion. Be sure to include a title on your display.

A science fair like this one is a great place to report your conclusions to people. You can see that all these people have created displays that let people know what they learned in their science experiment.

This young man is presenting his ideas to the class. As you present your conclusions, speak slowly and clearly and include facts that will keep people interested in what you have to say.

A Live Demonstration

On occasion your conclusions can be best communicated by giving a live **demonstration**. In the food experiment, you might set up small clear plastic cups with lemon juice, cooking oil, baking soda in water, salt water, and egg whites. Be sure to label each cup clearly.

When you have an audience, explain your experiment. Then, dip strips of red and blue litmus paper in each food. Blue litmus paper turns red when it touches acids. Red litmus paper turns blue when it is placed in alkaline foods. Place the dipped papers in front of each food's container. Tell your audience whether each food is acid or alkali. After the demonstration, tell the audience your conclusions.

What Did You Learn?

A valid conclusion is a summary based on facts. It takes into account all the information you collected and tells the final results. To communicate your conclusion, you will need to study all your data, write up your report, and present your ideas.

Do not forget that the purpose behind your experiment was to prove your hypothesis. Your hypothesis in the food experiment was that most foods were acidic. You tested 50 different foods found in most kitchens. You found that 31 of them had acid. One valid conclusion you can draw is that your hypothesis was correct.

Once ideas are proven right or wrong, scientists must take one more step. They must teach others what they discovered by reporting their conclusions.

Glossary

acids (A-sids) Sour-tasting matter that form salts when mixed with an alkali.

alkaline (AL-kuh-lun) Having to do with matter that forms salts when it mixes with acids.

communicate (kuh-MYOO-nih-kayt) To share facts or feelings.

data (DAY-tuh) Facts.

demonstration (deh-mun-STRAY-shun) A presentation to show how something works.

diagrams (DY-uh-gramz) Pictures of something.

graphs (GRAFS) Pictures that sort facts and make them clear.

hypothesis (hy-PAH-theh-sis) Something that is suggested to be true for the purpose of an experiment or argument.

information (in-fer-MAY-shun) Knowledge or facts.

litmus (LIT-mus) Powdery matter that turns red when it touches acids and blue when it touches alkalis.

neutral (NOO-trul) Neither acidic nor alkaline.

organize (OR-guh-nyz) To have things neat and in order.

summary (SUM-uh-ree) A short account of something that has been said or written.

valid (VA-lid) Correct or true.

visual (VIH-zhuh-wul) Having to do with sight; facts given in picture form.

volcano (vol-KAY-noh) An opening in the surface of Earth that sometimes shoots up a hot liquid rock called lava.

Index

Web Sites

Due to the changing nature of Internet links, PowerKids Press has developed an online list of Web sites related to the subject of this book. This site is updated regularly. Please use this link to access the list:
www.powerkidslinks.com/usi/repoconc/